AF688616

Recycle

A Practical Guide to Waste Management, Sustainable Living, and Recycling Techniques for Home, Business, and Environmental Impact

Noah Alexander Hayes

ISBN: 9781807655327

Published by TPSA Publishing

Table of Contents

Introduction

Recycling is the procedure of producing or manufacturing new products from a device that has originally fulfilled its purpose. If these used commodities are disposed of in a proper, environmentally friendly way, they can be used again in an effective way. The process of recycling has been used for things that are made from materials such as aluminum, plastic bottles, and certain kinds of paper that can be removed from the regular litter and put in an appropriate recycling bin. The composition of bio degradable waste such as food or garden waste is also chosen for recycling. Materials that are used for recycle are brought to a special collection center or picked up from the curbside, then set, cleaned, and convert into new materials required for manufacturing.

In other words, recycling of a material would produce a brand new supply of the same material for instance; used office paper trash would be reprocessed into new office paper, or used foamed polystyrene into new polystyrene. Anyhow, this is much difficult or too expensive. So "recycling" of many products or materials involves their reuse in producing different new materials. One more form of recycling is the restoration of certain materials from multiple products, either due to their true value, for example leads from car batteries, or gold from computer components etc. Critics misunderstood the economic and environmental profits of recycling over its costs, and advise that patrons of recycling sometime make matters worse and suffer from confirmation bias. Specially critics dispute that the costs and energy used in the collection and transportation process reduce (and outweigh) the costs and energy saved in the production process. Special jobs produced by the recycling industry can be a poor trade for the jobs lost in logging, mining, and other industries link with virgin production; and some materials such as paper pulp can only be recycled a few times before material degradation prevents further recycling.

Basic of recycling

Recycling as we know is the process of collecting and processing materials that would else ways be thrown away as litter and turning them into new products. Recycling can prove very beneficial for your community and the environment.

What is waste

Waste has been a very important environmental issue everywhere since the industrial revolution. Alongside the waste we create at home, school and at other public places like parks, and markets, it also include those waste that gather from hospitals, industries, farms and other sources. Humans rely on material things and at last, all material becomes waste.

Kinds of waste

Waste is those specific items that we (individuals, offices, schools, industries, hospitals) do not need and dispose off. Sometimes there are some things which we have that the law requires us to discard because they can be harmful for our health. Waste comes in innumerable sizes and much can be as small as an old toothbrush, or as large as the body of a school bus.

Everyone creates waste, somehow, many people are very environmentally conscious and they create very little waste. Likewise, some countries do a great job in creating less waste and managing the rest. While others countries are very horrible and have created huge environmental problems not only for the people but also for animals living there.

Throughout the world, communities handle their waste or garbage differently. Some common methods of managing their waste include land filling, recycling and composting. Other communities strongly commence waste reduction and litter prevention/control aimed at reducing the production of waste their first priority. Some communities also engage in waste-to-energy plants and unhealthy waste disposal programs. Usually waste could be liquid or solid waste. Both of them could be dangerous. Liquid and solid waste types can also be grouped into organic, reusable and recyclable waste.

Let us see some details below: Liquid Waste

Waste can come in liquid form and some solid waste can be transformed into a liquid waste form for disposal. It includes several point source and non-point source discharges such as storm water and wastewater. Some examples of liquid waste include wash water from homes, liquids used for cleaning in industries and waste detergents.

Solid Waste

Solid waste is a special form of garbage, refuse or rubbish that we make in our homes and at other places. These wastes include old car tires, old newspapers, broken furniture and even food waste. They may include any waste that is not liquid.

Dangerous Waste

Dangerous or harmful waste is those that specifically threaten public health or the environment. Such waste could be inflammable such as a waste that can easily caught fire, reactive waste that can easily explode, corrosive that can easily eat through metal or toxic that is poisonous to human and animals alike. In several countries, it is required by law to involve the appropriate authority to administer the disposal of such hazardous waste. Examples include fire extinguishers, old propane tanks, pesticides, mercury-containing equipment such as thermometer and lamps (e.g. fluorescent bulbs) and batteries.

Waste material that recycled

With the growing human population the needs for the people also increases. But the point of interest is that are there enough sufficient resources to service all your needs. What if these resources end, this is one thing we need to understand? We must start recycling waste to converse our natural resources. Recycling is commonly the process of reusing the items from which utility can still be derived. It is very important to recycle waste so that you can at least converse some of our natural resources for our generations to come. Numerous products such as paper, cardboards, and cups come from trees.

As a matter of fact trees are our natural wealth, you can converse trees by recycling the paper products or we can decrease the number of trees cut down a year. This is only one form of waste recycling. One should understand and know the importance of recycling waste materials. One clear benefit of recycling is it saves our resources. It will be wise enough to reuse metal item as metal reserves may be depleting. Important waste material that recycled includes things made of paper, metal, glass, electronic waste etc.

Paper waste

Different Paper waste items include newspapers, books, cardboard, magazines, boxes and envelopes.

Their collection

Many paper recycling companies collect material and the biggest task for them is probably the collection, transporting and sorting of the waste paper. This happened because we always mix paper to many other waste items and get them contaminated with food, plastics and many other different metals. Often collected paper is sent back to the junkyard because they are very contaminated for use. Always try to keep waste paper in independent grades at home or in the office, For instance, never mix newspapers and corrugated boxes up. When all paper get recovered, it is sent to the recycling center, where it is packed, graded, put into bales and sent to the paper mill after the long process. At the foundry, all the paper is stored in a stockroom until it is needed.

Screening

From the storage shelves, this paper moved into a large paper-grinding machine called a vat (pulper). In this place the paper is chopped into teeny pieces, mixed with water and chemicals and then heated up to break it down into organic plant material called fiber. Afterward, it is screened to remove contaminants such as different tiny pieces of plastic and globs of glue.

Deinking

This involves cleaning the pulp with chemicals to separate printing ink and glue residue. Often, a process called floatation is applied to remove tough stains and stickies. This process of floatation involves the use of chemicals and air to create bubbles which consume the stickies in the pulp.

Bleaching and refining

Distillation involves beating the recycled rubbish to make them ideal for again paper-making. After the refining, many additional chemicals are added to remove any dyes from the paper. It is then bleached to whiten and brighten it up.

Paper making

At this stage, the rubbish is ready to be used for paper. Sometimes new litter is added to give it additional strength and smoothness. Some water is added to the rubbish and sprayed onto a large metal screen in a continuous mode. The water is pooped on the screen and the fibers begin to link with each other. When it moves through the paper-making machines, press rollers squeeze out more water and then heat them dry and coat them up. Then this new processed paper sent for use.

Glass wastes

All products made of glass such as broken bottles, wine and beer bottles can be recycled in a refinery. Recycling of glass starts in the home, there is a solid reason why most of the local councils provide separate containers for green, brown, plain glass and even glass from broken windows. The reason behind that they are all made very differently and mixing them can create big problems at the recycling center.

Collection

Many cities have special collection spots. Trucks often pick glass waste from your home, or you might be required to drop them off at a special point in your town. In all cases, try to do what the main authorities have suggested. Always be sure, that you know the various glass types that are collected from your home. Always wash and separate them into the separate grades for collection.

Cleaning of Glass

The glass is transported to the processing plant through trucks. Where polluted material such as metal caps and plastic sleeves are removed, various grades are treated separately. Afterward clean glass is crushed into tiny pieces called cullet. Cullet is in high demand from the glass manufacturers. It melts at a very lower temperature and it is also cheaper than raw glass materials.

Ready for use This cullet is then transported to special glass-making factories. Here, it is mingled with sand, soda ash and limestone. It is heated at very high temperature and then melted into glass liquid. This liquid is then poured into specific moulds that give glass its special shape. This glass is used for many things, depending on from what grade they were recycled. Some important items made of recycled glass are fibre-glass, countertops, bottles and jars.

Aluminum waste

Aluminum cans are very lightweight, user friendly, handy, and keep liquor cold. They are often used for package soda, beer, and other drinks, and account for nearly all of the beverage packaging market for some products. Now, there has been immense progress in recycling aluminum cans. In 2003, Americans recycled more than 62.6 billion aluminum cans. Those cans, placed end-to-end, could make 171 circles around the earth. In every minute, an average of 105,800 aluminum cans is recycled. That is how important recycling of can has become.

But what is involved here? Here is it...

Collection of Aluminum cans Local councils arrange special can recycling containers that are clearly noticeable. This helps people to understand what to place in them. These Cans include soda, fruit and vegetable cans. Trucks come for these at pick up spots to the recycle centers. These cans may make of metallic or steel, but people do not know the difference very clearly.

Preparation

At the collection center, a huge magnet is rolled over them as they move on the conveyor belt to pick out all the necessary metal and steel cans. Only the aluminum cans are washed, crumble, and condensed in to 30-pounds briquettes for other companies for further processing. The rest is also sorted and sent to their appropriate recycling centers.

Melting of Material

The whole crushed cans then loaded into a burning furnace, where all printing and designs from the cans are removed, melted and blended with new (virgin) aluminum. This molten (liquid) aluminum is poured into special moulds and made into bars called ingots.

Special Sheets The block of metal then fed into powerful rollers, which crush them into special sheets of aluminum of about 25.4 in thickness. These thin sheets are rolled into coils and sold or sent to can-making factories. They use the special aluminum coils to prepare cans and containers for other food and drink manufacturers. It is predicted that cans collected at collection points take up to 60 days to be reappear in the shops as new cans containing your favorite soda, juice or food.

Recycling of Electronic Material

Innumerable people are recycling their electronic waste, with different companies involved in such recycling programmes observing an increase in the volume of recyclables collected.

According to the National Environment Agency (NEA), approximately 60,000 tones of such electronic waste are generating every year, equivalent to the weight of 160 Boeing 747 jets. Items constitute unwanted television sets, printers and computers. More than half of it comes from households and the rest from industry, such as information and communications technology equipment from industrial sectors.

Although electronic waste constitutes a large amount of total waste generated in Singapore annually, it is still very important to make sure that electronic waste is disposed of carefully because they contain certain dangerous components which can harm humans and the environment. Manycomputer makers companies are also collecting more electronic trash. Toshiba has specific collection points at its service centers for his customers to drop off their old laptops and batteries. It has reported an increase trend in the items collected. This is also the case for Dell, which is providing recycling services for his customers to donate or recycle their old IT equipment.

How is electronic waste recycled

Most of the electronic waste goes through a recycling system called a WEEE (Waste Electrical and Electronic Equipment), which not only recycles 95 to 98%, by weight, of all electronic waste passed through it, but it also ensures that any data left on hard drives and memories are thoroughly destroyed too.

Picking Shed, first of all, the each item is sorted by hand and batteries and copper and are extracted for quality control.

Initial Size Reduction Process, items are cut into pieces as small as 100mm to prepare the electronic waste to be thoroughly sorted. This is also where the data destruction takes place.

Secondary Size Reduction, the small debris is shaken to make sure that it is equally spread out on the conveyor belt, before it gets broken down even more. Any type of dust extracted is disposed of in an environmentally friendly way.

Overband Magnet, using magnets, steel and iron are also removed from the debris.

Metallic & Non-Metallic Content aluminium, copper and brass are separated from the non-metallic content. The metallic can then be reused and resold as raw materials.

Water Separation, water is used to separate plastic from the glass content. Once divided all raw materials can then be resold.

Recycling CRT Monitors and Televisions

Recycling of Cathode Ray Tubes, most commonly found in computer monitors and old televisions, is much more complicated than recycling most electronic waste. This is because toxins found within CRT's are among the most dangerous toxin and CRT's have lead in the glass which can seriously harm the environment and our health at the same time.

For the separation of Monitor Body and Cathode Ray Tube, firstly the front of the monitor is removed so that the tube can be removed easily from within. The shell is then recycled along with regular electronic waste.

Size reduction process, the tubes and screens are cut down into tiny pieces. Any glass dust created in this process is disposed of carefully, in an environmentally friendly way.

Metal Removal, the broken bits and some parts of glass are then passed under magnets, where iron and steel is extracted from the glass material. The remaining material is then passed through Eddy Currents, which removes any aluminium and copper. These materials are then collected and reused as raw materials.

Washing Line, the remaining glass is then cleared of oxides, phosphors and dust extracts, leaving just clean glass to be sorted.

Glass Sorting Line, the final stage of CRT recycling sorts the leaded glass from the unleaded glass. These can then both be used in the creation of new screens, which is called a closed loop recycling.

Food and Garden Waste

Food and garden waste currently make up more than 60 percent of average household litter yet it could be put to a range of domestic and commercial uses. It is estimated that we throw away one-third of all fruit and vegetables we buy. We can all take simple steps to reduce this waste. Start by buying only as much fresh produce as you need. If you are a business, donate your leftover food to a charity or food redistribution network.

How it is Collected

Basildon Council has provided every household with a special kitchen food caddy to help make recycling cooked and uncooked food waste more simple and hygienic. Simply empty your kitchen food caddy into your green bin along with any garden waste.

To ensure your green bin is collected make sure it is placed outside your property by 7.30am on the day of collection.

It Include

All cooked and uncooked food waste used at home

Peelings and leftover fruit used by members

Newspaper (but only if used as a liner in the kitchen caddy

Grass Cuttings

Shrubs

Plants

Branches

Weeds

Leaves

Cut Flowers

House Plants Recycling of Plastic Waste

How it is Recycled

The best thing that can happen to food is that it makes it to our plates and is enjoyed. Avoiding throwing out food that could have been eaten will save you money and help reduce greenhouse gas emissions. However some food waste is inevitable. Egg shells, banana skins and tea bags are never going to be on the menu. Home compositing is a great way to stop this sort of waste ending up in landfill, and our gardens will really thank us for it. If you live in an area that has a local food waste recycling collection service, you can use this to dispose of anything you do not want to eat, or compost at home. It can be recycled into a good quality soil improver or fertilizer and even generate electricity that can be fed back into the national gridor can be useful for you at home as well.

Vessel Compositing

This technique involves mixing food waste with garden waste and shredding it and then composting it in an enclosed system for around 2-4 weeks (temperatures of up to 70°C speed up the process and ensure any harmful microbes are killed off). The material is then left outside to mature for a further 1-3 months with regular turning and checks to ensure quality before going on to be used as soil conditioner.

Anaerobic Digestion

Another technique of recycling of food waste is the use of microorganisms to break down food waste, animal manure, slurries and energy crops in the absence of oxygen, inside an enclosed system. As it breaks down it gives off methane, which is collected and converted into biogas and used to generate electricity, heat or transport fuels. It also creates a nutrient-rich digestate that can be used as a fertilizer for agriculture and in land regeneration.

Its impact on Environment

Reducing food waste is a major issue and not just about good food going to waste; wasting food costs the average family with children almost £60 a month and has serious environmental implications too.

The amount of food we throw away is a waste of resources. Just think about all the energy, water and packaging used in food production, transportation and storage. This all goes to waste when we throw away perfectly good food.

Cheese is a good example, feeding and milking the cows, cooling and transporting the milk, processing it in to cheese, packing it, getting it to the shops, keeping it at the right temperature all the time. If it then gets thrown away it will most likely end up in a landfill site, where, rather than harmlessly decomposing as many people think, it rots and actually releases methane, a powerful greenhouse gas.

Tips for controlling food Waste

Never forget to make the most of your food and drink and try to avoid wasting food in and make it priority. For ideas, recipes and simple tips visit Love Food Hate Waste. You could also try to compost at home.

A food waste small caddy in your kitchen can help you to separate out your food waste for recycling and composting. This can be emptied into your compost bin or council food waste bins every couple of days.

Your council may recommend that you should line your food waste caddy with a liner or newspaper. Only use liners that are recommended by your council as some may not break down in the composting process.

Where possible keep your bins out of direct sunlight and keep the bin lid closed.

How Plastic Waste Recycled

Mostly plastic jugs and bottles are hundred percent recyclable. The Plastic #1 and #2 are the most commonly used forms of plastic in bottled water containers and the most commonly collected. The plastic in jugs and bottles is weightless durable, flexible and, yes, at times controversial. A modern study show that mixed plastics packaging (trays, tubs, pots, films etc) can be mechanically recycled and is both economically and environmentally effective to do so. Some additional research and work is being conducted in this area and it is hoped that the UK will have comprehensive infrastructure for the collection, sorting and reprocessing of these valuable resources in the future. Now almost, 49% of councils collect other rigid plastic packaging such as pots, tubs and trays in household recycling collections.

The Plastic household containers are generally marked with a number that indicates the type of plastic they are made of. Consumers can then use this information to find out whether or not certain plastic types are collected for recycling in their area. It is important to remember that these codes were not developed to communicate recyclability, but rather to communicate the unique qualities of the different types of plastic. To be sure of a jug or bottle's recyclability in your area, jump to the recycling locator.

How it Recycled

For the recycling of plastic many methods have been used by companies. They first sort out the material and then shredded them into tiny pieces. Then the material is washed and melted into large factories.

sorted

shredded

washed

melted

pelletised

It is a two-fold Process

sorting is mainly done automatically with a manual sort to ensure all contaminates have been removed

plastic is either melted down directly and moulded into a new shape, or shredded into flakes then melted down before being processed into granulates.

Its Impacts on Environment

Plastic is a popular, useful material but we use a lot of it. By optimizing the lifespan of plastics and by reusing and recycling items as many times as possible, for example, by recycling used plastic bottles into new ones, we can reduce our need to create new plastic. This means we can:

Conserve non-renewable fossil fuels

Reduce the consumption of energy used

Reduce the amount of solid waste going to landfill

Reduce the emission of gases like carbon dioxide into the atmosphere.

Product made from recycled Plastic

There is a wide range of products made from recycled plastic, including:

polyethylene bin liners and carrier bags

plastic bottles

flooring and window frames

building insulation board

video and compact disc cassette cases

fencing and garden furniture

water butts, garden sheds and composters

seed trays

fleeces

Degradable plastics are not best choice for biological waste treatments like anaerobic digestion or composting and are not likely to break down in a garbage site because of the lack of UV light and oxygen. In addition, they could potentially affect the quality of recycled plastic if they enter the plastic recycling system in high concentrations. There are also several concerns found about the fragmentation of the material into small particles, which can remain in the environment for months.

A recent study recommended energy from waste as the best disposal option for degradable plastic.

Bioplastics

A special term is given to a wider group of plastics that can be either biodegradable, bio-based, or both.

Biodegradable plastics

This plastic should not be confused with degradable plastic; they are generally derived from renewable material that can be broken down into simple compounds such as carbon dioxide and water, by naturally occurring microorganisms. The term 'biodegradable' often does not imply any particular time scale or environment for biodegradation to occur, nor the extent.

Compostable plastics

There are a subset of biodegradable plastics, which means they will biodegrade and disintegrate during composting to produce carbon dioxide, water, inorganic compounds and biomass at a rate consistent with other known compostable materials. They leave no visually distinguishable or toxic residues. The European standard for compostability is the EN 13432 and relates to industrial composting conditions only.

Recycling of Polystyrene Waste

When thrown out, polystyrene does not break down and takes up an important space in landfill. Polystyrene bags can easily blown by the wind, ending up as litter in the environment where it can harm animals and marine life at the same time.

Polystyrene cannot be recycled from your kerbside recycling bin. But now you can help the environment and recycle polystyrene. Residents and small business can now drop-off expanded polystyrene for recycling at locations across the globe. Once collected, your polystyrene will be recycled and re-manufactured into domestic building products of your use. Despite the growing attractiveness of polystyrene, municipalities and organizations are facing huge problem in disposal of polystyrene packaging and products. Being large and bulky, polystyrene take up significant space in rubbish bins which means that bins becomes full more quickly and therefore needs to be emptied more often. Polystyrene is lightweight compared to its volume so it occupies lots of precious landfill space and can be blown around and cause a nuisance in the surrounding areas.

Recycle process

There seems to be a common misconception found among people that polystyrene is non-recyclable. Being a thermoplastic, it can actually be melted and molded very easily into many different plastic items. Currently the recycling of polystyrene (or EPS foam) basically follows the following process:

Segregation process

EPS foam products are separated from other wastes and then sorted into big area.

Compaction process After the segregated EPS foam, these products are fed to a compactor in order to reduce its volume. Some compactor systems have a compaction ratio of up to 50:1, which means that it can reduce the volume by up to 98%.

Shredding Process

Larger pieces are shredded into special flakes. Packaging "peanuts small EPS foam pieces used to cushion fragile items, normally skip this step and are fed directly to the pelletizing machine.

Melting process

The flakes are forced through pelletizing extruders where they are heated and melted, then allowed to cool in order to solidify into new shapes. The resulting material can then be used, through reheating and melting, to produce clothes hangers, picture frames, DVD cases and numerous other plastic products.

Major Bottlenecks

Although the Alliance of Foam Packaging Recyclers have reported that the recycling rate for post-consumer and post-commercial EPS in the United States have risen to 28% in 2010 from around 20% in 2008, this value is still lower than most solid wastes. According to USEPA, auto batteries, steel cans and glass containers have recycle rates of 96.2%, 70.6% and 34.2% respectively. Because it is bulky, EPS foam takes up storage space and costs more to transport and yet yields only a small amount of polystyrene for re-use or remolding (in fact, polystyrene accounts for only 2% of the volume of UN compacted EPS foams). This provides little incentive for recyclers to consider EPS recycling.

Products that have been used to hold or store food should be thoroughly cleaned for hygienic reasons, thus compounding the costs. For the same reasons, these products cannot be recycled to produce the same food containers but rather are used for non-food plastic products. The manufacture of food containers, therefore, always requires new polystyrene. At present, it is more economical to produce new EPS foam products than to recycle it, and manufacturers would rather have the higher quality of fresh polystyrene over the recycled one.

Costly process

The cost of transporting bulky polystyrene waste discourages recyclers from recycling it. Organizations that receive a large amount of EPS foam (especially in packaging) can invest in a compactor that will reduce the volume of the products. Recyclers will pay more for the compacted product so the investment can be recovered relatively easier.

There are also breakthroughs in studies concerning EPS recycling although most of these are still in the research or pilot stage. Several studies have found that the bacteria Pseudomonas putida is able to convert polystyrene to a more biodegradable plastic. The process of polystyrene depolymerization – converting polystyrene back to its styrene monomer – is also gaining ground.

Meanwhile, for the rest of us, we can start reducing our polystyrene consumption by opting to use products that can be reused, such as bringing our own coffee mugs and food containers to stores that serve their food and drinks in EPS foam. A small change in our lifestyles can make a big difference for the environment.

Other Things

There is a long list of other material that can be recycled. Organic waste can be composted and turned into fertilizer. Rubber tires can be shredded, decontaminated and made into insulation or other new innovative products. If you are looking for new ways to recycle, simply give a moment's thought when you throw something out. Could it be reused or broken down in a useful way?

Important Recycling Techniques

There are various ways through which materials can be recycled. The special technique which is used to create the new materials from the old rubbish depends on what the type of material is. Here is a list of different techniques:

Concrete Recycling process

Battery Recycling process

Biodegradable Waste Recycling process

Electronic Waste Recycling process

Printer & Ink Cartridges etc

Glass recycling

Paper recycling

Ferrous Materials recycling

Non-Ferrous Materials recycling

Plastic recycling

Ship breaking recycling

Textile Recycling

Recycling facts

More than 20,000,000 Hershey's Kisses are wrapped each day, using 133 square miles of tinfoil. All that foil is recyclable, but not many people realize it.

One-third of the water used in most homes is flushed down the toilet. You can recycle newspaper and white paper, glass bottles and, aluminum cans, steel cans or plastic containers with a 1 or 2 on the bottom.

An aluminum can that is thrown away will still be a can 500 years from now.

Your ID card will last approximately 450-650 years.

The energy which saved from recycling of one aluminum will operate a computer for three hours.

Recycling of one aluminum can save energy to run a TV for 3 hours-or the equivalent of a half gallon of gasoline.

We use 80,000,000 aluminum soda cans every year.

If all our newspaper was recycled, we could save about 250,000,000 trees each year according to some sources.

Recycling a stack of newspapers just 3 feet high can save one tree

We throw away enough office paper annually to build a wall twelve feet high stretching from Los Angeles to New York City.

Americans use 2,500,000 plastic bottles every hour. Most of them are thrown away.

The energy saved from recycling one glass bottle can operate a 100-watt light bulb for 4 hours.

Every month we throw away enough glass bottles and jars to fill up a giant skyscraper

Glass can be recycled an infinite amount of times and can used.

Each of us generates on average 4.4 pounds of waste per day per person.

Did you know that the state of Maryland generated enough municipal solid waste in 2000 to build a wall 3′ wide x 6′ high and 3,839 miles long? That is long enough to go from Baltimore to Miami over 3.5 times.

Americans make more than 200 million tons of garbage each year, enough to fill Cole Field house from top to bottom twice a day

Importance of recycling

Recycling is one of the best ways for everyone to have a positive impact on the world in which we all live. There are many reasons why it is important to recycle and compost as much as possible because it not only conserves raw materials but also making new products out of recycled materials which reduce the need to consume precious resources. So recycling helps in the protection of raw materials or also natural habitats for the future.

Recycling is important for both the natural environment and us. We must act as fast as the amount of waste we create.

The amount of trash we create is gradually increasing because:

Increasing of wealth means that people are buying more products and therefore creating more waste.

Increasing population means that there are more people on the earth to create waste.

Many new packaging and technological products are being developed, and many products contain materials that are not biodegradable.

New lifestyle has adopted, such as eating fast food, means that we create additional waste that isn’t biodegradable.

Environmental Importance

For environment recycling is very important as waste has a huge negative impact on the natural environment.

Many harmful chemicals and greenhouse gasses that are released from rubbish in landfill sites. Recycling helps to reduce this pollution caused by waste.

Habitat destruction and global warming are some affects caused by deforestation. Recycling reduces the need for raw materials so that the rainforests can be preserved.

Huge amounts of energy are used when making products from raw materials. Recycling requires much less energy and therefore helps to preserve natural resources.

Importance of Recycling for people

Recycling is important to cities around the world and for people living in them.

No more space can found for waste. Our landfill sites are filling up fast, by 2010, almost all landfills in the UK will be full.

Reduce financial expenditure in the economy. Making products from raw materials costs much more than if they were made from recycled products.

Preserve natural resources for future generations. Recycling reduces the need for raw materials; it also uses less energy, therefore preserving natural resources for the future.

Recycling is important for earth safety

Recycling of different products helps the environment. For example, we know that paper comes from trees and many trees are being cut down just to produce paper. By recycling it, we can help in decreasing the number of trees that are cut down. Products made from raw materials that came from our natural resources should be recycled so that we can help preserve the environment.

Recycling is important for Energy

it takes few energy to process recycled materials than to process virgin materials. For example, it takes a lot less energy to recycle paper than to create a new paper from trees. The energy from transporting virgin materials from the source is also saved. Saving energy also has its own benefits like decreasing pollution. This creates less stress on own health and our economy.

Recycling Helps Mitigate Global Warming

For saving energy the recycling process prove very helpful. The greenhouse from factories and industrial plants are lessened and the use of fuels that transmit harmful gasses during production is also minimized. Recycling non-biodegradable waste (rather than burning it) will contribute a lot to help reduce air pollution and greenhouse gasses that depletes the ozone layer.

Recycling Reduces Waste Products

Junkyards are mostly composed of non-biodegradable waste which takes long time to decompose. By recycling, we can lessen the waste materials that are placed into landfills and we are able to make the most out of these materials. If we don't recycle, more and more garbage will go to landfills until they all get filled up. If that happens, where will the rubbish be placed? How would you like a land fill in your backyard?

There are also many companies out there that help reduce the problem of overfilling landfills by offering options (like ink cartridges) that make sense. When consumers seek out companies that have established green practices, like Carrot Ink, the consumer wins by not only helping the environment but also by saving money.

Recycling Save Money Recycling helps in many ways to save money. You can sell recyclable materials to organizations that are willing to buy it. Using products that are recycled lessens expenses. Products that are made from recycled materials are less expensive than products made from fresh materials.

At home, you can recycle biodegradable waste like eggshells, vegetable and fruit peelings and use them to fertilize plants. By doing a little research and getting creative you can save money and trips to the market while being kind to the planet.

As the population of the world increases recycling is becoming increasingly more important. Our technologically advanced societies are creating more and products and packaging that look good and are indestructible, but can take centuries to break down.

In order to combat the rise of factors that are produced by non-environmentally conscious groups, it is up to the growing numbers of individuals and companies that want to inhabit a healthier planet to make a difference.

Judicial recycling use of Resources

Recycling also promotes judicial and renewable use of resources. This process ensures that there is no discriminate use of any material when available in plenty in the present. Recycling is encouraged at all levels, from school to government offices and even at international levels. This means we can preserve all important resources for our future generation, without any compromise in the present.

Recycling is useful for natural resources

If we will not recycled old and used materials, the new products will made from extracting fresh raw materials from beneath the earth through mining and extraction. Recycling prove very helpful in conserving important raw materials and protects natural habitats for the future. Conserving natural resources such as wood, water and minerals ensures its optimum use.

Recycling is helpful for jobs

Recycling is very useful for the environment and apart from that it also creates green jobs for people. According to the U.S. Bureau of Labor Statistics, green goods and services accounted for 3.1 million jobs in the United States in 2010.

Useful Tips for Recycling

As the planet's resources are increasingly fewer it is becoming more and more important to recycle. Rather than your waste going in to landfill and harming the environment, it can be recycled and reused even at home very easily. However, it can be difficult to recycle as much as you can and should. These easy tips will help you to recycle more in your home.

Separate Your Trash

One of the main reasons that people do not recycle as much as they could at home because they do not know how to sort their trash. Once you have finished with an item or a piece of packaging, think about whether or not it can be recycled at home. If you are not sure, many packaging items will state on them if they can be recycled. If you can recycle it, give it a wash out if necessary, and then separate it. Better to keep a few separate containers in which you can put your recycling. You could even sort it into different material types, such as plastics, paper and card, and metal.

Recycle Machine

If you have a garden at your home, you can recycle you plant matter in order to make great compost. Not only will you be saving money on buying compost, you will also be reducing the amount of waste that goes into landfill from your home. You will need to get yourself a composter, but the rest is very easy and logical. When you have any garden waste, such as tree trimmings or mowed grass, you can put that into your composter. You can also add any green kitchen waste, such as vegetable peel and leftover uncooked fruit or vegetables.

Electrical Recycled

More items than you might think can be recycled. There are even places where you can recycle old electrical items and gadgets even at home easily. So, if you have upgraded your cell phone or your MP3 player has worn out, then think about recycling them, rather than throwing them out or letting them sit at the back of your drawer. Look out for charities that collect old and broken personal electronics for recycling. Then you are not just helping the planet, you are also supporting a good cause, satisfactory isn't it.

Textiles Recycling

If you have any old clothes or textile items that can be recycled at home, use them in your recycling compost. You can also look for a local recycling bank that accepts textiles. If you have any clothes that are in good condition, then you could take them to a charity shop to be recycled or sold on. If you have some clothes that are if really poor condition, you can recycle them yourself by cutting them up and using them as rags for household cleaning.

Reuse of Material

Another way of recycling wastes is to reuse your material. This is actually more environmentally friendly, as a lot of energy is used to process recycling. There are many things you can reuse that you might otherwise throw away. Use old jam jars or spaghetti sauce jars as storage pasta, rice, or flour. Use old plastic ice cream or margarine tubs to take sandwiches to work or to store leftovers in the fridge. Be creative and you'll be surprised at how much you can reuse.

In order to halt climate change, we are all going to have to live more sustainable lifestyles. Recycling may seem like a small step, but every little helps. In fact, if every home in America switched just one roll of virgin toilet roll to 100 percent recycled then 544,000 trees would be saved from being cut down. So, just a small amount of recycling might make more of an impact than you'd expect.

Recycle Bins

A recycling bin is a special container used to hold recyclables waste before they are taken to recycling centers. These containers you might have seen outside your home, in parks or streets and are used to store recyclable material. Recycling bins are available in various sizes for use in homes, offices, and large public facilities. Separate containers are often provided for paper, tin or aluminum cans, and glass or plastic bottles at public places.

Many recycling bins are designed to be recognizable easily, and are marked with special slogans promoting recycling on a blue or green background along with the universal recycling symbol. Others are intentionally unobtrusive. Bins are sometimes of different colors so that user may differentiate easily between the types of materials to be placed in them. While there is no universal standard, the color blue is widely used to point out a bin is for recycling in public settings.

Recycling bins are a common element of municipal kerbside collection programs and In order to make sure that more and more people participate in recycling activity local municipal corporations are now encouraging owners of home to dispose off their waste products in recycle bins. These bins are given to the home owners by municipal corporations themselves. These are normally called household bin are most commonly type of bin today.

Recycle bins is a great way to start recycling. It has helped awfully in last few decades to process waste products which otherwise would have ended up in filling up the garbage. Recycling in other way would help in Energy Conservation, Reducing pollution and landfill reduction.

Large bins are also placed at major parks, malls and offices so that people throw away their waste products at appropriate places. Offices dispose large quantities of paper daily while malls dispose off large quantity of disposable glasses and plates each and every day.

Few countries in the world have proper recycling colors and logos to help people identify the most appropriate bin. This helps people to put proper things in proper places so that recycling firms can easily recognize different items from the large quantities of waste products.

Importance of Recycle Bins

Mostly people understand that recycling is important, not everyone takes the time to do it, Because In many cases, this is happened due to the places where they live and work is not set up in a manner that makes it convenient to dispose of items that can be recycled in an appropriate manner. If it is not easy for people to stow unwanted items in containers designated for recycling, they will often simply throw them in the garbage , even though they know better. Most of the time, people are simply going to go with the trash disposal option that is the easiest and most expedient solution at the time rather than searching for the best long term solution.

That is why it is important to understand the importance of recycling bins and need to make it first priority to make decisions about where to place containers for recycling and what types are necessary. It is not difficult to place recycle bins and containers in strategic locations throughout your living and/or working space. You simply need to think about what types of waste are likely to be generated and locate appropriate containers for recycling in the areas where people are most likely to need them.

By spending a little money now to purchase recycle bins and containers, you can make great strides in reducing the carbon footprint of your lifestyle, as well as the environmental impact of those who live and work with you. This is the best first step to take if you want the people you live and work with to develop the habit of recycling rather than throwing away items that are recyclable.

Different Types of Recycling Bins

There are three important types of recycling bins that have been provided by the government to dispose of the recyclable items. Many special containers are designedto gather items for recycling. Before making a decision about the best bin options for your home or office, find out what types of recycling are available in your local area and learn the guidelines for how items need to be sorted prior to being picked up or dropped off at your local recycling center.Each color of the bins represents different uses. It is also important to factor information about the volume and kinds of waste produced in your home and office before selecting bins and making decision about quantity and placement. This information will help you to determine how many containers you need and what instructions need to be provided about placing items in them. These recycling bins come in different sizes – 80 liters, 120 liters, 240 liters, 360 liters, and 660 liters etc.

While some types of recyclable items are common in mostly homes and offices, there is some variation. Common recycling bin needs include:

Aluminum

Batteries

Cardboard

Electronic equipment

Leaves and grass clippings

Toner and ink cartridges

Motor oil

Paper

Plastic bottles

Plastic gift cards

Steel

Other recyclable post consumer waste items

Important container features

Try to select bins that are color coded so that it will be easy for people to remember what types of waste go in each container. First of all, you are likely to need to label the containers to make it easy for people to learn what types of waste go in each container, but with time, if you are consistent, it will become easy for people to associate particular items with certain colors.

Special size

Select containers that are large enough to be able to go for several days without being emptied, yet that are not too large to be easily carried to the curbside for pickup or placed in a vehicle to be transported to your local recycling center.

Stackable - If you'll be placing several bins in a single room - such as containers for plastic bottles and aluminum cans in the kitchen it's probably best to go with stackable bins. This will allow you to make the most of the valuable floor or cabinet space in the room.

Where to Place

Do remember that many people are not likely to take the time to recycle if it is not convenient for them to do so. No matter what types of containers you have, they will only be effective if people use them. Containers dedicated to gathering paper for recycling should be placed in or near areas where paper waste is generated, such as by copy machines, printers, desks where bills are paid, and other appropriate places.

Safety is also an important consideration when starting a business or household recycling initiative. When you are purchasing containers to hold paper for recycling, for example, it's also advisable to invest in a paper shredder. Use the shredder to destroy any documents that could provide ammunition for identity thieves prior to placing them in the bin. Additionally, if you are recycling containers that held food or beverage products, be sure to rinse them thoroughly before setting them aside for recycling. This will help ensure that your efforts to be environmentally responsible don't result in a pest infestation in your home.

Kerbside rubbish Bin

A kerbside rubbish bin is the bin with a red lid and black body and is collected weekly by the authorities.

Residents will have the option of three rubbish bin sizes, 80 litre, 120 litre and 240 litre.

Council encourages residents to choose the smallest bin size especially for their needs (larger bin sizes incur greater fees).

Material

This service is for the collection of residual waste from around the home that cannot be reused, recycled or composted, and includes:

General household waste

Plastic bags

Nappies

Broken drinking glasses or broken window glass

What should not place

Hot ashes or liquids

Chemicals

Large heavy objects

Car batteries

Recyclable material (glass bottles and jars, paper, aluminium cans etc)

Compostable material (lawn clippings, vegetable scraps, weeds, leaves etc)

All of Greater household rubbish has been taken to Cosgrove garbage area, with no further processing to separate out any recyclable material. Do the right thing and use the right bin.

If it exceeds 80 kilograms in weight it will not be collected by authorities. Please remove material prior to the next collection.

Yellow Recycle Bin

This special yellow kerbside recycling bin is the bin with a yellow lid and blue body and is collected on alternate fortnights to green organics, the same day your garbage is collected.

Material place

Glass bottles and jars, beer bottles etc

Milk and juice cartons

Steel or aluminum cans

Newspaper, magazines, mixed paper and cardboard trash

Plastic containers used at home.

Some recyclable plastic containers are identified with special symbol normally located at the bottom of the container.

Polyethylene Terephthalate, used for soft drink bottles
High Density Polyethylene, used for milk bottles, cordial and juice bottles
Unplasticised Polyvinyl Chloride, used for clear cordial and juice bottles
Low Density Polyethylene, used for ice cream lids, garbage bags
Polypropylene, used specially for ice cream containers, yoghurt

pots
Polystyrene, used for yoghurt pots plastic cutlery, hot drink cups, take away containers Miscellaneous, most often polycarbonate PC i.e. used for water cooler bottles, sunglasses, DVDs, iPhone and computer cases, signs and displays, certain food containers, nylon

Material that should not place in

Plastic bags Hot ashes or liquids Compostable material (lawn clippings, vegetable scraps, weeds, leaves etc) Polystyrene Car batteries Engine oil Scrap steel and aluminium Clothing and old toys Electronic equipment (TVs, electrical components etc.) Broken drinking glasses or broken window glass

Recycling process

After your recycling has been collected, it is transported to the Visy materials recovery facility in, Australia, where it is processed and recyclable materials are separated from any contaminants. Plastic bags supplied from the supermarket cannot be placed in this recycle bin. Do not fill plastic bags with recyclables and put it in the recycle bin as plastic bags cannot be recycled.

If your bin is exceeding 80 kilograms in weight it will not be collected. Please remove material prior to the next collection.

Green Recycle Bin

Green waste is the material we all generate in our kitchens and around our gardens and comprises items such as vegetable scraps, lawn clippings, small woody prunings and weeds. This special kerbside green organics bin is the bin with a lime green lid and dark green body and is collected on alternate fortnights to recycling, the same day as your domestic garbage bin.

Special Features

This service is for the collection of green organic waste from the garden and kitchen. The bin size is 120 or 240 liter and these organics bins will be collected fortnightly on the alternate fortnight to your recycle bin.

What can Place In

Lawn clippings Weeds Vegetable scraps Loose branches (trimmed to fit inside the green waste bin with the lid closed) Leaves Tealeaves and Coffee Grounds

What should not place in

All other materials which is Garbage Plastic bags and plant pots Paper, cardboard and plastics Glass and metal Soil and stone Meat and bones Nappies

How to use

Your green organics service is for the collection of lawn clippings, prunings, weeds, vegetable scraps and the like. In some circumstances your bin may become odorous. These conditions may arise when air is not able to flow through the contents of the bin, for example if the material is very wet and water accumulates in the bottom of the bin. Although the bins are vented, the air flow may not be sufficient to remove all moisture from the contents.

The following tips may help the householders to avoid these conditions:

Try to cut lawns when the grass is dry at the end of the day in the evening.

If possible schedule your garden maintenance close to your collection day. For instance if you cut your grass fortnightly, time it so that lawn clippings are only in your bin for a few days

Is it possible to stockpile your garden waste and load the bin just before collection day?

Place woody type material in the base of the bin so that any accumulated moisture will not become odorous.

Loosen the contents of the bin occasionally to aerate material and also to break up grass clippings that have become crusty. Wet grass clippings can form a crust that prevents contents leaving the bin during the collection process

What happen after Collection

Your green organic material is taken to Western Composting Technology where it is processed into a high grade compost soil conditioner known as GV Compost, available at local nurseries. If it is up to exceeding 80 kilograms in weight it will not be collected. Please remove material prior to the next collection.

Special Public Place Bins

Some Public place bins are located all around Greater Shepparton in parks, gardens and commonly used public areas. There are two types of bins - Waste (including doggie-bag bins) and Recycling. These bins are to be used in the same way as your bins at home, with the red topped bin being for waste and the yellow topped one for recycling.

Public place waste bins are for use by the public only. Businesses must not use public place waste bins for the disposal of commercial or shopkeeper waste.

Recycling waste Baskets

These special recycling & Waste Baskets are best for office recycling use. Its efficient and beautiful design fits snugly beside any desk and can be used for collecting recyclables or waste from the office.

Its Feature

Can be used for trash or recycling

One injection molded seamless piece

Available in a wide variety of colors

Holds the Hanging Waste Basket (BC1500)

Made with 35% recycled content

Custom stamping available

Lightweight, yet durable

Fits beside or under most desks

What should place in

Paper waste

Cans and bottles

Disposable cups and plates

Clips, pins etc

What should not Place in

Unnecessary waste of Glass material

Fruits and vegetable trash

Jars and glass bottles

These have the capacity of 5.8 quart and its dimension is 12.25" H x 8.25"W x 14.25"D.It is specially made from ammonium.

A two in one Recycler Basket

This special 2-in-1 Recycler makes recycling very easy for people. This special bin allows you to either contain trash and recyclables all in one place or sort recyclables. The above unit is perfect for collecting recyclables and has a special fold away handle that makes it easy to take out and empty into your recycling bin. The lower bin part tilts out for easy access, and holds a 7 gallon trash bag in place with the Liner Lock™ system. Alternatively, you can use its lower bin to sort another type of recyclable and carry it out with the built-in handles.

Contain trash and recyclables all in one location or sort your recyclable in separate parts

Sleek design blends into kitchen décor easily

Features Liner Lock™ bag management in lower bin

Easily removable lid

Two Fold-away handles

5.8 gal sorting bin

7.4 gal tilt out bin with Liner Lock™ handles

Total capacity up to 13.2 gal

A Perfect item for multi-stream or single stream recycling.

Special Bins in public areas

Some public areas such as parks have special litter bins which are placed alongside paths frequently walked by visitors to encourage them to throw their waste in them. This also encourages public not to create waste because it can create an unhealthy and aesthetically unpleasant social environment. These Public trash bins are accompanied by recycling bins or "bottle bins" to be used for recycling glass and plastic. Some public bins, especially in "green" restaurants, accept organic waste to be composted.

Bins in outdoor locations or other busy public areas are usually mounted to the ground or wall. This discourages theft, and also reduces vandalism by making it harder for someone to physically move or maneuver the bins; to spill their contents on the ground for example, or to use the bin as an effective weapon to hit people over the back with.

Recycling Centers

A special Heart Lake Community Recycling Centre, located in the City of Brampton at 420 Railside, is now open for people to help them in the recycling process. Drop-off your household hazardous waste, garbage and items for donation or recycling process in this centre and help government in keeping the environment clean.

Special Services

Special Public Drop-off Area for Recyclable and Non-Recyclable Materials

Special household Hazardous Waste Drop-off Area and Reuse Centre

Business Hazardous Waste (Participating CRCs only, conditions apply)

Special reusable Goods Drop-off

Special reusable Goods Stores (at specified sites).

Special Items

Backyard composters

Compost (loose or in bulk)

Garbage tags

Rain barrels

Recycling boxes

Green bins and kitchen containers

Caution

asbestos (PDF, 2 pages, 880 KB)

hazardous waste from businesses

institutional, commercial or industrial hazardous waste

Recycling Center in Hampshire

The Recycling Center is located at 147 Hampshire Street in the back right corner of the Public Works' yard. We are open on Tuesday and on Thursday from 4:00pm - 7:30pm and on Saturdays from 9:00am - 4:00pm.

Location

The yard is closed to the public at all other times. The Center is open to Cambridge residents only and small businesses and non-profit organizations in Cambridge with up to 50 employees. Use of the Center is free of charge. Materials that small businesses/non-profits can bring are marked by special sign.

Items that accepted

Batteries (non-alkaline only) accepted

Books (3-ring binders in good condition OK, Resident s are welcome to take up to 10 books per visit.) are accept by the authorities.

Bottles & Cans*

Cardboard Boxes*

CDs, DVDs, Video Games (in original packaging)

Cell Phones

Electronics and Small Appliances (Residents only. Any small appliance with a cord or battery operated device that one person can lift. No computer monitors, no TVs. Large appliances must be scheduled for curbside pickup.)

Flags

Food Scraps (residents only)

Metal Items* (no sealed containers or large appliances)

Mercury Items (fluorescent bulbs*, thermometers, thermostats)

Packing Peanuts (no block Styrofoam, no cellulose)

Paper*

Plastic Bags*

Plastics, Large Item laundry baskets, crates, toys, etc.

String Lights

Toner & Ink Cartridges

Item that are not accepted

No Floppy Disks, VHS or Cassette Tapes

No Car Batteries (bring to junk yard)

No Chemicals

No Computer Monitors or TVs

No Construction Debris

No Furniture

No Hazardous Waste

No Paint

No Trash

No Yard Waste or Wood

No Window Glass

Recycling Centers in Perth

There are 9 staffed Recycling Centers in Perth and Kinross for people to drop their items for recyclable process. These centers are found in Aberfeldy, Auchterarder, Bankfoot, Blairgowrie, Crieff, Friarton (Perth), Inveralmond (Perth), Kinross and Pitlochry. In early winter 2012, the new Inver almond Recycling Centre has been opened for the public. Please have a look at the special list below for Recycling Centre opening hours and types of material accepted.

Aberfeldy Recycling Centre

Material Accepted

Batteries (household), Bicycles, Cans (including aluminum foil, aerosols, biscuit tins), Car Batteries

Bulky Goods,

Cardboard,

Cooking Oil,

Electrical & Electronic Equipment,

Engine Oil,

Fridges/Freezers, Fluorescent Tubes

Garden Waste,

General Waste,

Glass,

Long Life Light Bulbs,

Paper, Plastic (rigid plastic packaging),
Rubble Stone and Soil,

Scrap Metal,

Telephone Directories, Textiles (clothes and shoes),

Timber, (hand) Tools for Reuse,

Tyres (maximum two per visit, strictly householders only) LPG Gas Bottles should be returned to the supplier.

Opening Hours

All year Tuesdays and Thursdays 9.00am - 7.00pm

Saturdays and Sundays 10.00am - 5.00pm

Auchterarder Recycling Centre, the Public Park, Western Road,

This very famous recycling centre accept material of wide variety such as,

Batteries (household), Bicycles, Cans (including aluminium foil, aerosols, biscuit tins), Car Batteries

Bulky Goods,

Cardboard,

Cooking Oil,

Electrical & Electronic Equipment,

Engine Oil,

Fridges/Freezers,

Garden Waste, General Waste,

Glass,

Paper,

Plastic (rigid plastic packaging),

Rubble Stone and Soil,

Scrap metal,

Telephone Directories,

Textiles (clothes and shoes),

Timber, (hand) Tools for reuse,

Tyres (maximum two per visit)

LPG Gas Bottles should be returned to the supplier.

A range of companies are happy to accept them for reuse, for free.

Recycling facilities in Paisley

Household waste and recycling centers are large facilities provided by government, where you can recycle most of rubbish of your house. The refurbishment at the Underwood Road centre is now complete and the site has re-opened for people. Small recycling points provide large bins for combinations of glass, newspapers, cans, mixed plastics, clothes and textiles.

Several other recycling points include our new range of bin units. These units combine collection bins for various types of waste to meet local demand.

Items that Accepted

Glass

Cans

Food Carton

Green waste

White Goods

Paper and Cardboard

Redfield Road Household Waste and Recycling Centre

There is a specific facility in Nottingham where you can take your household recycling, garden waste and waste. This site is called a Household Waste and Recycling Centre. Household waste is the waste you would normally put into your garbage bin at home, grass cuttings from your garden and items of furniture.

Ideally you can also recycle and compost as much of your household waste as possible at home. Before you go the Household Waste and Recycling Centre, think first about what you can separate out for home recycling and composting.

Item that accepted

Batteries

Car batteries

Flat glass

Florescent tubes

Food tins and drinks cans

Fridges and freezers

Garden Waste

Gas Bottles

Hardcore and rubble

Large appliances

Mixed glass, bottles and Jars

Mixed paper and card

Mixed textiles and clothes

Mobile phones

Plastics

Printer cartridge

Scrap metal

An important thing to consider is that you cannot dispose office waste in that centre.

Recycling is Important for Environment

Recycling helps the environment in many different ways. It means lots more trees and trees help to clean the air we breathe in which means more oxygen and fresh air. Trees are also act as house for animals like birds and squirrels. Trees can also play important role in preventing mudslides from happening by keeping the dirt stronger. One of the most important reasons why trees are important is that they give us foods and other natural resources.
Recycling keep environment Clean Another way recycling helps the environment is that recycling keeps the local landfills or dumps from growing. That means cleaner neighborhoods. You don't want a dirty town or neighborhood, right so the question should ask daily is, do you want earth to be covered in garbage? I definitely don't want the earth to be covered with litter. What I am trying to say is recycling helps keep the environment and earth clean.

Recycling stop Pollution

If we recycle, we won't have to manufacture things as much. Recycling help people in disposing off their waste easily or it help in saving energy because you probably don't realize that when we manufacture things, we sometimes pollute the air. Polluting the air does not help the environment, it hurts it. We don't want this to happen. Polluting the air hurts our lungs, we won't be able to breathe, and we might die. Recycling helps us and our environment. Recycling Saves Animals Another reason why recycling helps the environment is that it protects animals. Recycling helps animals in the environment by keeping habitats clean and safe. If we don't recycle, the fish or other animals, like turtles, might get stuck in plastic six-pack soda can holders and eventually die. The world would not be the same without animals. Can you imagine the world without animals? Our world definitely would not be the same without animals. Recycling helps animals and the environment. Recycling definitely helps the environment and our future. The future of the world depends on recycling. If we recycle, earth will have a brighter future. When earth has a brighter future, we will too. I don't know about you, but I want the world to have a great future. Let's do our job and recycle. Let's save our earth.

Recycling Helps Global Warming

Recycling is an easy way to effectively stop global warming. Recycling is to render untill suitable for reuse in case you did not know. Global warming is the rising in temperature of the world. Recycling is something that everyone should do and if many people do it global warming will stop.

Less Landfills

The first major advantage of recycling is that if we recycle less waste will go into landfills. A garbage area is a site where waste materials are buried. What happens once trash goes into landfills is that it gets covered in dirt and some of it decomposes over time, but water and moisture can filter through the remaining waste and pick up metals, minerals, organic chemicals, bacteria, viruses, and other toxic chemicals and end up running into our ground and surface water and polluting it. Did you know that the average American produces about 4.4 pounds of trash (that trash is not recycled) every day?

Less Methane and Incinerators

Second main advantage is that if we recycle less methane will be released into the air from landfills. Methane is colorless, odorless, highly flammable gas and it is also linked to global warming. When it is not thrown into landfills it is put in an incinerator which is poorly sited and unpopular. Even worse they are so energy-inefficient they use more energy than they create. Only after those two options are worn out do we use recycling and to stop global warming recycling waste needs to be a first option.

Less Recycling Aluminum Cans

The final advantage is that if you recycle one aluminum can that aluminum can has enough energy to power one T.V. for three hours without releasing all of the harmful chemicals a T.V. powered by raw materials would.

It also takes 95% less energy to recycle one aluminum can than to make three hours T.V. from raw materials. Aluminum can also be recycled over and over again without decreasing its price. Saving manufacturers money as well as saving you money while helping the environment as well. If you recycle aluminum cans you are helping the cause of stopping global warming.

Recycling helps in making pollution free environment

Recycling aluminum cans and paper causes significantly less water pollution than the manufacturing of products from virgin materials. Landfills contain an abundance of toxic materials. Over time, these materials seep pollutants into groundwater, a major source of drinking water for many people. Manufacturing products using recycled materials consumes considerably less energy than the industrial processes used to create products from virgin resources, thus drastically reducing greenhouse gas emissions. Landfills are also responsible for producing methane, a potent greenhouse gas, as well as known carcinogens and other air pollutants. The process of incineration releases toxins into the air along with heavy metals and unknown chemical compounds that are a result of incomplete combustion. How to encourage recycling Recycling is becoming increasingly important as the whole world is becoming more aware of environmental issues. More offices, schools and businesses are "going green," finding ways to embrace the environment by recycling. The following steps will help you encourage your co-workers and bosses to recycle at work.

Some instruction

Make it simple to recycle. This includes contacting your local sanitation company to acquire latest recycle bins. Get as many different kinds as you can; plain paper, aluminum, cardboard, plastic and newsprint. Set these up in areas of the office those are readily accessible and logical. For instance, the aluminum bin would go in the kitchen area while the plain paper one would go near the copier or printers.

Set up a special shredder in the mail or copier area so employees can shred their unwanted papers in it. Then you can use that papers in the shipping department use the as packing material.

Check with your local Bureau of Waste Prevention or local sanitation department to invite one of their staff members to talk to your organization about the latest techniques. Not only will you get invaluable tips for recycling, but these people will help motivate your employees or co-workers to help the environment. Once people get the mindset of how easy recycling is, they will be more willing to do it.

Create a contest between departments to see who can recycle the most. For example, have personnel from each department choose something their department needs, then agree that they can receive the item (within reason) for their department if they win your contest. Or have all departments bring in items that are highly recyclable that an office supply organization pays for (such as printer cartridges or cell phones); then have an office party with the money raised by this recycling drive.

Ask management to purchase environmentally-friendly light bulbs, plumbing fixtures such as low-flow toilets and recycled products such as paper and printer cartridges to encourage people to see and practice recycling both as consumers and producers.

Make colorful posters showing all the items that are recyclable-paper, toner cartridges, aluminum cans, soda bottles, cardboard and even computer hardware-to educate people to keep remember that do not throw their waste and reuse it.

Tips for making environment clean

Some useful tips for making our environment more clean and healthy are as follow,

Recycled glass reduces related air pollution by 20 percent and related water pollution by 50 percent. If it isn't recycled it can take a million years to decompose.

By the time a child is toilet trained, a parent will change between 5,000 and 8,000 diapers, adding up to approximately 3.5 million tons of waste in U.S. landfills each year. Whether you choose cloth or a more environmentally-friendly disposable, you're making a choice that has a much gentler impact on our planet.

American businesses throw away 21 million tons of paper every year, equal to 175 pounds per office worker. For a quick and easy way to halve this, set your printer's default option to print double-sided (duplex printing). And when you're finished with your documents, don't forget to take them to the recycling bin.

There are 63 million newspapers printed each day in the U.S. Of these, 44 million, or about 69%, of them will be thrown away. Recycling just the Sunday papers would save more than half a million trees every week.

Nearly 90% of plastic water bottles are not recycled, instead taking thousands of years to decompose. Buy a reusable container and fill it with tap water, a great choice for the environment, your wallet, and possibly your health. The EPA's standards for tap water are more stringent than the FDA's standards for bottled water.

It's good for the air, the land, can shade your house and save on cooling (plant on the west side of your home), and they can also improve the value of your property. Make it meaningful for the whole family and plant a tree every year for each member.

Consider buying items from a second-hand store. Toys, bicycles, roller blades, and other age and size-specific items are quickly outgrown. Second hand stores often sell these items in excellent condition since they are used for such a short period of time, and will generally buy them back when you no longer need them.

The average cell phone lasts around 18 months, which means 130 million phones will be retired each year. If they go into landfills, the phones and their batteries introduce toxic substances into our environment. There are plenty of reputable programs where you can recycle your phone, many which benefit noble causes.

Wire hangers are generally made of steel, which is often not accepted by some recycling programs. So what do you do with them? Most dry cleaners will accept them back to reuse or recycle.

By maintaining your vehicle, you are not only extending the life of your vehicle, but you are creating less pollution and saving gas. A properly maintained vehicle, clean air filters, and inflated tires can greatly improve your vehicle's performance. And it might not hurt to clean out the trunk; all that extra weight could be costing you at the pump.

Each year the U.S. uses 84 billion plastic bags, a significant portion of the 500 billion used worldwide. They are not biodegradable, and are making their way into our oceans, and subsequently, the food chain. Stronger, reusable bags are an inexpensive and readily available option.

Each year 15 billion batteries produced and sold and most of them are disposable alkaline batteries. Only a fraction of those are recycled. Buy a charger and a few sets of rechargeable batteries. Although it requires an upfront investment, it is one that should pay off in no time. And on Christmas morning when all the stores are closed? You will be fully stocked.

An average individual should try to participate in their municipal recycling program, getting rid of mixed paper by that means rather than through a waste collection program. Yes, it means a little bit of sorting, but something recycling is better than, and more often than not, paper is dealt with aside from other sources of waste, such as bills or taxes, etc.

Disadvantages of Recycling

Recycling is not always cost-effective. Sometimes, there may be a need to establish separate factories to process reusable products. This may create more pollution in terms of cleaning, storage and transportation. Recycled Products May not Last for Long: Recycled products are always not of durable quality. Such items are mostly made of trashed waste, picked up from heaps other waste products which are of fragile or overly used. For this reason, recycled products are cheap and last for a shorter period.

Unsafe and Unhygienic Recycling Sites: Recycling sites are often unsafe and unhygienic. Places where all sorts of waste is dumped are conducive for debris formation and spread of disease and other dangers caused by harmful chemicals and waste. This not only causes widespread pollution but is harmful for dedicated people who recycle such products. Such waste if mixed with water, leads to leachate formation and leads to toxication of water bodies including drinking water.

Not widespread on Large Scale: Although recycling is an important step to minimize pollution, unfortunately this process is just a small part of long-term success. Recycling often occurs at a small scale-homes or schools and has failed to be useful at a large level such as at industries or holistically at a global stage. Saving paper at schools cannot be compared to oil spills or massive tree felling at industrial level.

Setting up new recycling unit involves high cost. This huge cost can come up as a part of acquiring different utility vehicles, upgrading the processing facility, educating residents by organizing seminars and other programs, disposing of existing waste and chemicals etc.

After weighing, the disadvantages of recycling, one can wisely take crucial steps involved in this process. Understanding the impact of recycling is essential on a large-scale which if done effectively can bring in massive positive results, beneficial to mutual existence of human beings and environment.

Conclusion

Upon reviewing thoroughly the recycling process, one can easily understand it's important for making environment clean .There are many ways in which recycling is in effective for everyone. We can confidently state that recycling is a profitable measure. In addition to these findings, we can also conclude that recycling would gain more profit if consumers had something to gain from recycling. Recycling offer various benefits to people, most important it helps in reducing pollution which is very essential for better health.

However government can also educate people by giving incentives like creating more recycling programs, raising awareness on the need for recycling to benefit our polluted environment, and most importantly the institution of bottle deposits across the nation. This monetary bottle deposit would be the key incentive to encourage recycling. Supported by the evidence that states with bottle deposits have higher recycling rates However very few countries of world are participating in the recycling, granting the other countries a more economical reason to recycle which will help maintains resources and reduces pollution in a drastic way. Thus, a greater number of recyclable products will be transformed into post-recycled goods. They will be available for use again, prolonging the natural resources that Earth will not be able to sustain indefinitely if we do not recycle. So it should be our first duty to save our environment by reusing our resources.

www.ingramcontent.com/pod-product-compliance
Lightning Source LLC
LaVergne TN
LVHW022318200926
848481LV00004B/1371
9781807655327